INDUCTION AND INTUITION
IN SCIENTIFIC THOUGHT

MEMOIRS OF THE

AMERICAN PHILOSOPHICAL SOCIETY

Held at Philadelphia

For Promoting Useful Knowledge

VOLUME 75

INDUCTION AND
INTUITION IN
SCIENTIFIC THOUGHT,

PETER BRIAN MEDAWAR
Director, National Institute for Medical Research, London

Jayne Lectures for 1968

AMERICAN PHILOSOPHICAL SOCIETY
INDEPENDENCE SQUARE ● PHILADELPHIA
1969

The Jayne Lectures of the American Philosophical Society honor the memory of Henry La Barre Jayne, 1857-1920, a distinguished citizen of Philadelphia and an honored member of the Society. They perpetuate in this respect the aims of the American Society for the Extension of University Teaching, in which Mr. Jayne was deeply interested. When in 1946 this organization was dissolved, having in large measure fulfilled its immediate purposes, its funds were transferred to the American Philosophical Society, which agreed to use them "for the promotion of university teaching, including *inter alia* lectures, publications and research in the fields of science, literature, and the arts."

Accepting this responsibility, the Society initiated in 1961 a series of lectures to be given annually or biennially by outstanding scholars, scientists, and artists, and to be published in book form by the Society. The lectures are presented at various cultural institutions of Philadelphia. Thus far the following, including the series published in the present volume, have been presented:

February 21, 28, March 7, 14, 1961. Per Jacobsson. *The Market Economy in the World of Today.* University Museum, University of Pennsylvania. Memoirs of the American Philosophical Society, Vol. 55 (1961).

March 7, 14, 21, 1962. George Wells Beadle. *Genetics and Modern Biology.* University Museum, University of Pennsylvania. Memoirs of the American Philosophical Society, Vol 57 (1963).

March 6, 13, 20, 1963. Doris Mary Stenton. *English Justice Between the Norman Conquest and the Great Charter, 1066-1215.* University Museum, University of Pennsylvania. Memoirs of the American Philosophical Society,Vol. 60 (1964).

March 10, 17, 24, 1964, Ellis Kirkham Waterhouse. *Three Decades of British Art: 1740-1770.* Philadelphia Museum of Art. Memoirs of the American Philosophical Society, Vol. 63 (1965).

May 3, 4, 6, 7, 1965. William A. Fowler. *Nuclear Astrophysics.* The Franklin Institute. Memoirs of the American Philosophical Society, Vol. 67 (1967).

February 14, 21, 28, March 7, 1966. Jacob Viner. *The Role of Providence in the Social Order: an Essay in Intellectual History.* University Museum, University of Pennsylvania.

October 31, November 7, 14, 1967. Douglas Bush. *Pagan Myth and Christian Tradition in English Poetry: Three Phases.* Free Library of Philadelphia. Memoirs of the American Philosophical Society, Vol. 73 (1968).

April 4, 9, 11, 1968. Sir Peter Medawar. *Induction and Intuition in Scientific Thought.* University Museum, University of Pennsylvania. Memoirs of the American Philosophical Society, Vol. 75 (1969).

Reprinted 1975.

FOREWORD

How do scientists' minds work as they try by observation, experiment, and reflection to solve the problems set for them by Nature? Is there an established method of scientific thinking? As Sir Peter Medawar has pointed out in the following pages, few scientists have attempted to analyze their own thought processes. They get on with their work and arrive at their conclusions without deeply considering the mental pathways they followed. Those who have attempted to analyze the scientists' method of thinking, from Francis Bacon and John Stuart Mill to twentieth-century writers, have mostly not been scientific investigators but philosophers and logicians. Working scientists have generally regarded with indifference the attempts of philosophers to categorize the steps of scientific thought, for example sharply contrasting "deduction" and "induction," for they do not recognize these processes as distinguishable in their own work.

Sir Peter Medawar, in his Jayne Lectures for 1968, brings to this important topic an unusual combination of scientific experience and philosophical reflection. Winner of the Nobel Prize for Medicine in 1960 for his researches on growth, aging, immunity, and cellular transformations, and author of several recent volumes on the philosophy of science, he sees the problem with clear eyes and arrives at an explanation of scientific thought—the "hypothetico-deductive" process, as he calls it—which scientific investigators will find truly descriptive of their manner of thinking as they go about their research.

Among his audiences at the University of Pennsylvania in April, 1968, there were not only scientists but historians of science, all of whom agreed that Sir Peter's exposition of the problem was the clearest they had ever heard. The American Philosophical Society, whose interests embrace both science and logic, is proud to have supported these lectures and to make them available to the scholarly public.

GEORGE W. CORNER

AUTHOR'S PREFACE

Induction and Intuition in Scientific Thought is a long title for so small a book, but I found no way of defining its subject matter more succinctly. These Lectures began in my mind in the form of a question: why are most scientists completely indifferent to—even contemptuous of—scientific methodology? Put generally, the answer could only be "because what passes for scientific methodology is a misrepresentation of what scientists do or ought to do." I therefore thought it important to explain what is wrong with the traditional methodology of "inductive" reasoning, as I see it, and to show that the alternative scheme of reasoning associated with the names of Whewell and Peirce and Popper can give the scientist a certain limited but useful insight into the way he thinks.

Like most scientists, I never write on subjects outside my own unless I am expressly asked to do so. I am therefore specially grateful to the American Philosophical Society for having invited me to deliver the Jayne Lectures in Philadelphia in the spring. The lectures are printed in the form in which I delivered them, with references, explanations, and digressions added in the form of notes.

PETER B. MEDAWAR

CONTENTS

INDUCTION AND INTUITION IN
SCIENTIFIC THOUGHT

I. THE PROBLEM STATED

1

I T IS NOT at all usual for scientists to deliver formal lectures on the nature of scientific method, particularly if they are still engaged in scientific research. Of course, it is an understood thing that scientists of a specially elevated kind, e.g. theoretical physicists, may from time to time express quietly authoritative opinions on the conduct of scientific enquiry, while the rest of us listen in respectful silence; but that a biologist should speak up where so many physicists and chemists have chosen to remain silent must seem to you to be yet another symptom of the decay of values and the loss, in this modern world, of all sense of the fitness of things.

Yet—if the task of scientific methodology is to piece together an account of what scientists actually *do,* then the testimony of biologists should be heard with specially close attention. Biologists work very close to the frontier between bewilderment and understanding. Biology is complex, messy and richly various, like real life; it travels faster nowadays than physics or chemistry (which is just as well, since it has so much farther to go), and it travels nearer to the ground. It should therefore give us a specially direct and immediate insight into science in the making. The wisest judgments on scientific method ever made by a working scientist were indeed those of a great biologist, Claude Bernard.[1]

We all know in rough outline what lawyers do, or clergymen, physicians, accountants, and civil servants; we have a vague idea of the codes of practice they must abide by if they are to succeed in their professional duties, and if we were to learn more about them we should be edified, no doubt, but not surprised. But what are scientists like as professional men, and how do they set about to enlarge our understanding of the world around us? There seems to be no one answer. The layman's interpretation of scientific practice contains two elements which seem to be unrelated and all but impossible to reconcile. In the one conception the scientist is a discoverer, an innovator, an adventurer into the domain of what is not yet known or not yet understood. Such a man must be speculative, surely, at least in the sense of being able to envisage what *might* happen or what could be true. In the other conception the scientist is a critical man, a skeptic, hard to satisfy; a questioner of received beliefs. Scientists (in this second view) are men of facts and not of fancies, and science is antithetical to, perhaps even an antidote to, imaginative activity in all its forms.

Let me begin with the scientist as a questioner of received beliefs. During the seventeenth century, when the new science came in on a spring tide,[2] and again during the nineteenth century, the forward movement of

[1] *Introduction à l'étude de la médecine expérimentale* (Paris, 1865), a work that suffers in translation (which may account for its limited influence in the English-speaking world).

[2] The phrase is Henry Power's: p. 192 of *Experimental Philosophy*, (London, 1644, Johnson Reprint Corporation, New York, 1966).

science called for a vigorous shaking off of scholastic constraints and religious superstition. No single work displays science in its critical temper more clearly than Francis Galton's *Statistical Inquiries into the Efficacy of Prayer,* published by the *Fortnightly Review* in its issue of August 1, 1872.

A belief in the efficacy of prayer (Galton reasoned) is something we all grow up with: it has behind it the formidable authority of habit, doctrine, and popular assent. But are there in fact any "scientific" grounds for supposing that prayers are answered: that what is prayed for comes about as a consequence of an act of prayer? One line of enquiry that seemed to Galton to promise a definite answer turned upon the health and longevity of the Queen and other members of the royal family—something prayed for weekly or even daily on a national scale, and sung for too, though in an imperative rather than a supplicatory mood. Do members of royal families live any longer as a result of these exertions of prayer on their behalf? Table 1, transcribed from Galton's paper, shows that if anything they fare worse than people of humbler birth.

The amplitude and frequency of prayers for the royal family cannot be assumed to be proportional to their sincerity, so Galton put the same question in a different way. No one can doubt the sincerity of prayers that appeal for the lives of newborn children: are then still-births any less frequent among the children of the devout than among the professional classes generally? Apparently not: Galton studied the number of still-births announced in *The Record* (a clerical newspaper) and in *The Times,* and found them to stand in exactly

TABLE 1

MEAN AGE ATTAINED BY MALES OF VARIOUS CLASSES WHO HAD
SURVIVED THEIR THIRTIETH YEAR, FROM 1758 TO 1843.
DEATHS BY ACCIDENT OR VIOLENCE EXCLUDED.

			Average	Eminent Men [1]
Members of Royal houses....	97	in number	64.04	
Clergy	945	"	69.49	66.42
Lawyers	294	"	68.14	66.51
Medical Profession	244	"	67.31	67.07
English aristocracy	1,179	"	67.31	
Gentry	1,632	"	70.22	
Trade and commerce........	513	"	68.74	
Officers in the Royal Navy...	366	"	68.40	
English literature and science.	395	"	67.55	65.22
Officers of the Army	569	"	67.07	
Fine Arts	239	"	65.96	64.74

[1] The eminent men are those whose lives are recorded in Alexander Chalmers' *General Biographical Dictionary* (32 vols., London, 1812-1817) with some additions from the *Annual Register.*

the same proportion to the total number of recorded deaths. The data are shaky, of course, and Galton was quite aware of the shortcomings of his analysis. His purpose was above all to show that such an analysis can in fact be done.

Galton's most telling argument was founded upon the policy of insurance companies in fixing the rates of annuities. To buy an annuity is to pay a capital sum at (for example) retirement, in return for which the company undertakes to provide the investor with an annual income until he dies. The rates offered by different companies are competitive and must be judiciously worked out, for if the annuitant lives beyond the calculated expectation the insurance company will be out of pocket. This being so,

It would be most unwise, from a business point of view, to allow the devout, supposing their greater longevity even probable, to obtain annuities at the same low rates as the profane. Before insurance offices accept a life, they make confidential inquiries into the antecedents of the applicant. But such a question has never been heard of as, "Does he habitually use family prayers and private devotions?" Insurance offices, so wakeful to sanatory influences, absolutely ignore prayer as one of them. The same is true for insurances of all descriptions, as those connected with fire, ships, lightning, hail, accidental death, and cattle sickness. How is it possible to explain why Quakers, who are most devout and most shrewd men of business, have ignored these considerations, except on the ground that they do not really believe in what they and others freely assert about the efficacy of prayer?

I have not done justice to the range and analytical skill of Galton's polished and urbane analysis, and I shall have done him a positive injustice if I leave you with the impression that he was merely having a go at religious belief. The rhetorical force of his argument would have been greatly weakened if it had been crudely irreligious. Prayer, he tells us, may strengthen the resolution and bring serenity in distress; it is an appeal for help; Galton did not "profess to throw light on the question of how far it is possible for man to commune in his heart with God." His reasoning was thus "scientific" in the territory in which he exercised it, but also in the territory he disclaimed.

Reasoning in this style is by no means confined to or even specially characteristic of scientific enquiry. Galton's first step was to assume the truth of an opinion for which there was a certain obvious *prima facie* case, namely that what is prayed for may come about through prayer; then he examined some of the logical consequences of holding that opinion; then, thirdly, he took steps to find out whether or not those logical expectations were indeed fulfilled. The argument was made out by reasoning, not by asseveration; the matters of fact upon which the judgment turned were, if not known, then knowable by everyone; and the testimony of inner voices went unheard. His great achievement was, of course, methodological. He brought within the domain of science matters until then thought to lie outside its competence: "the efficacy of prayer seems to me . . . a perfectly appropriate and legitimate subject of scientific inquiry." The reasoning may be empirically wrong, but it is not fallacious; an answer will be arrived at by this style of reasoning or not at all.

The critical task of science is not complete and never will be, for it is the merest truism that we do not abandon mythologies and superstitions but merely substitute new variants for old. No one of Galton's stature has conducted a statistical enquiry into the efficacy of psychoanalytic treatment. If such a thing were done, might it not show that the therapeutic pretensions of psychoanalysis were not borne out by what it actually achieved? It was perhaps a premonition of what the results of such an enquiry might be that has led modern psychoanalysts to dismiss as somewhat vulgar the idea that the chief purpose of psychoanalytic treatment is to effect a cure. No: its purpose is rather to give the patient a new and deeper understanding of himself and of the nature of his relationship to his fellow men. So interpreted, psychoanalysis is best thought of as a secular substitute for prayer. Like prayer, it is conducted

in the form of a duologue, and like prayer (if prayer is
to bring comfort and refreshment) it requires an act of
personal surrender, though in this case to a professional
and stipendiary god.

Nor has anyone yet conducted a formal analysis of the
all but universal belief that dreams are messages of
some kind; that dreams convey significant information
clothed in a dark and ancient symbolism which only the
initiated can decode. Analysis, I suspect, would reveal
that dreams, whatever else they may be, are not messages or communications of any kind. The utter non-
sensicality of dreams—their glorious emancipation from
the confinements of time and place and cause and
sense—is probably the most significant thing about
them, the property from which the student of mind
has most to learn. If these newer inquiries were to be
set in train, and were to have the outcome I have pre-
dicted, the resentment and sense of outrage they would
give rise to would be indistinguishable in character and
psychological origin from that which exploded nearly
one hundred years ago over Galton's analysis of prayer.

2

The layman's conception of the scientist as a critic,
a skeptic, a man intolerant or contemptuous of conven-
tional beliefs, is obviously incomplete. The exposure
and castigation of error does not propel science forward,
though it may clear a number of obstacles from its path.
To prove that pigs cannot fly is not to devise a machine
that does so. To explode the myth of the Chimera

makes it no easier to transplant a kidney from (say) ape to man.

The layman sees the other profile too. A scientist is a man who weighs the earth and ascertains the temperature of the sun; he destroys matter and invents new forms of matter, and one day he will invent new forms of life. But how has he achieved the understanding that makes this possible? What methods of enquiry apply with equal efficacy to atoms and stars and genes? What *is* "The Scientific Method"? What goes on in the head when scientific discoveries are made?

Rhetorical questions: and when we try to answer them a remarkable state of affairs is revealed. The scholarly discipline that might be expected to hold the answers is unpopular and in the main, in its larger ambitions, unsuccessful. If the purpose of scientific methodology is to prescribe or expound a system of enquiry or even a code of practice for scientific behavior, then scientists seem to be able to get on very well without it. Most scientists receive no tuition in scientific method, but those who have been instructed perform no better as scientists than those who have not. Of what other branch of learning can it be said that it gives its proficients no advantage; that it need not be taught or, if taught, need not be learned?

It will not do to say that a scientist learns by apprenticeship, implying that he learns to do his own work by studying the Works of others, for scientific "papers" in the form in which they are communicated to learned journals are notorious for misrepresenting the processes of thought that led to whatever discoveries they describe. The scientist is not in fact conscious of acting out a method. If a scientist is more or less successful in the enterprise he is engaged on, he attributes it to having enjoyed more or less of luck or learning or perceptiveness or flair, *never* to the use or misuse of a formal methodology. How very unlike a deductive exercise, such as we carry out when trying to derive a geometric theorem; here, if something goes wrong, our first thought is that we have made a logical (that is, a methodological) mistake.

Of course, the fact that scientists do not consciously practice a formal methodology is very poor evidence that no such methodology exists. It could be said—has been said—that there is a distinctive methodology of science which scientists practice unwittingly, like the chap in Molière who found that all his life, unknowingly, he had been speaking prose. Yet it may be revealing that not one of those whom we recognize as great methodologists of science was a practicing scientist himself. Francis Bacon was a lawyer and a man of affairs; a sociologist of science, if you like, and (if you like) a playwright. John Stuart Mill was a deeply learned and humane man, a political theorist and a sociologist in the modern sense, but though his "strong relish for accurate classification" had been gratified by lectures and books on botany and zoology,[3] his deeper scientific knowledge

---

[3] J. S. Mill, *Autobiography* (London, 1873). Mill attended lectures on zoology in Montpellier in 1820, and there seems no doubt that his thought on methodology was strongly influenced by the study of a subject overwhelmed by a multitude of "facts" which had not yet been disciplined by a unifying theory. Coleridge described it as "notorious" that zoology had been "falling abroad, weighed down and crushed as it were by the inordinate number and multiplicity of facts and phenomena apparently separate, without evincing the least promise of systematizing itself by any inward combination of its parts" (*General Introduction*, or *A preliminary Treatise on Method: Encyclopaedia Metropolitana* [London, 1818]).

came at second hand from William Whewell's *History of the Inductive Sciences* (1837). Whewell himself did not practice science nor add to it, except by way of nomenclature, but he was deeply enough informed about all its branches to have become the outstanding methodologist of his day.[4] Karl Pearson was a mathematician; Stanley Jevons and John Maynard Keynes were economists; C. S. Peirce was, as Karl Popper is, a great philosopher. Why did not scientists come forward and expound their own methodology? One did so: Claude Bernard, whom I have already mentioned; but his opinions seem to have made so little impact on the English speaking world that his name is mentioned in only two of a dozen well-known texts on scientific methodology in my shelves.

Unfortunately, a scientist's account of his own intellectual procedures is often untrustworthy. "If you want to find out anything from the theoretical physicists about the methods they use," said Albert Einstein, "I advise you to stick closely to one principle: don't listen to their words, fix your attention on their deeds."[5] Darwin's case is notorious. In his autobiographical

---

[4] In *The Philosophy of the Inductive Sciences* (London, 1837). For Whewell as polymath and nomenclator, refer to E. W. Strong, *Jour. Hist. Ideas* **16** (1955): p. 209; P. J. Wexler, *Notes and Queries*, n.s., **8** (1961): p. 27; S. Ross, *Notes and Records of the Royal Society* **16** (1961): p. 187. Among the familiar words he invented are *anode, cathode, ion, anion, cation, eocene, miocene, pliocene,* and of course *physicist* and *scientist.* Earlier variants of the latter cited in the *Oxford Dictionary of English Etymology* (ed. C. T. Onions) are *sciencer, scientiate, sciencist,* and *scientman.*

[5] "On the Method of Theoretical Physics," in *The World as I See It* (London, 1935).

sketch,[6] contemporary with the sixth edition of *The Origin of Species,* he said of himself that he "worked on true Baconian principles, and without any theory collected facts on a wholesale scale" (p. 83); but later in the same work (p. 103) he said that he could not resist forming a hypothesis on every subject, and he gave away his true opinions (as opposed to the opinions which he felt became him) in letters to Henry Fawcett and H. W. Bates.

Darwin's self-deception is one that nearly all scientists practice, for they are not in the habit of thinking about matters of methodological policy. Ask a scientist what he conceives the scientific method to be, and he will adopt an expression that is at once solemn and shifty-eyed: solemn, because he feels he ought to declare an opinion; shifty-eyed, because he is wondering how to conceal the fact that he has no opinion to declare. If taunted he would probably mumble something about "Induction" and "Establishing the Laws of Nature," but if anyone working in a laboratory professed to be trying to establish Laws of Nature by induction we should begin to think he was overdue for leave.

[6] *The Life and Letters of Charles Darwin,* ed. F. Darwin (London, 1887). The letters to Fawcett and to Bates are in *More Letters of Charles Darwin,* eds. F. Darwin and A. C. Seward (London, 1903), pp. 176, 195. To Fawcett he wrote (18 September, 1861):

"About thirty years ago there was much talk that geologists ought only to observe and not theorize; and I well remember someone saying that at this rate a man might as well go into a gravel-pit and count the pebbles and describe the colours. How odd it is that anyone should not see that all observation must be for or against some view if it is to be of any service."

To Bates (22 November, 1860):

"I have an old belief that a good observer really means a good theorist."

You must admit that this adds up to an extraordinary
state of affairs. Science, broadly considered, is incom-
parably the most successful enterprise human beings
have ever engaged upon; yet the methodology that has
presumably made it so, when propounded by learned
laymen, is not attended to by scientists, and when
propounded by scientists is a misrepresentation of what
they do. Only a minority of scientists have received
instruction in scientific methodology, and those that
have done so seem no better off.

One way out of this dilemma is to argue that scientific
methodology is understood intuitively by scientists and
needs to be propounded only for the benefit of other
people. Nearly all scientists are loud in deploring the
utterly unscientific way in which everyone else carries
on—politicians, educationalists, administrators, sociologists—and it is upon *them* that they urge the adoption
of the scientific method, whatever it may be. John
Stuart Mill, the most influential of all methodologists,
was certainly not trying to teach scientists their business. On the contrary, his ambition was to analyze and
expound their methods in the hope that the complex and
baffling problems of society would eventually give way
before their use. In a sense this was Bacon's ambition
too, for though one cannot be confident of any simplified
interpretation of that brilliant and strangely compounded character, yet his *New Atlantis* is the very con-
summation of what he thought the application of his
methods might achieve.

Perhaps then we should no longer think of scientific
methodology as a discipline of which the chief purpose
is to teach scientists how to conduct their business, but

rather as an attempt to get non-scientists to pull themselves together and smarten up and generally speaking be much more scientific than they are. Many modern methodological texts have therefore a strong orientation towards the social and behavioral sciences, as if sociologists and social anthropologists were backward because (poor things) they had not been properly brought up in the manners and usages of polite science. While I respect this evangelistic mission, I am not in sympathy with it. The "backwardness" of sociology (as in the nineteenth century of biology) has little now to do with a failure to use authenticated methods of scientific research in trying to solve its manifold problems. It is due above all else to the sheer complexity of those problems. I very much doubt whether a methodology based on the intellectual practices of physicists and biologists (supposing that methodology to be sound) would be of any great use to sociologists. On the contrary, the influence of inductivism, the subject of my next lecture, has in the main been mischievous. It has stirred up in some sociologists the ambition to ascertain the laws of social change, above all by the painstaking accumulation of data out of which general principles will in due course take shape. The elevated prose and studied postures of a flourishing school of social anthropology in France today are best explained away as a reaction against the crude scientism of those who have urged upon sociologists the adoption of a style of investigation which they do not use themselves and cannot authenticate from their own experience.

3

I have said so much that is critical of scientific methodology that you may wonder why I should have chosen to lecture upon it at all. I seem to do nothing but find fault.

If I have given that impression I must at once correct it. Even if it were never possible to formulate *the* scientific method, perhaps because there is no such thing, yet scientific methodology, as a discipline, would still have a number of distinctive and important functions to perform. For in the practice and interpretation of science a number of real problems arise which are common to all sciences but are "formal" in the sense that they do not depend on what the particular sciences are about. These are ample agenda for a school of methodology: let me now quickly mention three.

1. The problem of *validation:* of the grounds upon which general statements may be judged true or false or merely probable, and of the methods by which we may quantify their degree of imprecision. Under this heading I classify the illuminating developments of modern statistical analysis, particularly in the domain of small-sample theory, so much of it the work of mathematicians turned scientist or of mathematically minded biologists.[7] Matters of validation are important in the experimental sciences, but not as important as they are sometimes made out to be. (I shall argue in my next lecture that an obsessional preoccupation with matters to do with ascertainment is part of the heritage of the inductivism.) It is in the *generation* of scientific knowl-

[7] E.g. R. A. Fisher, F. W. Yates, "Student," J. H. Gaddum.

edge, not in its interpretation or in a retrospective analysis of "the data," that scientists are oppressed by the fear of error. It is a truism to say that a "good" experiment is precisely that which spares us the exertion of thinking: the better it is, the less we have to worry about its interpretation, about what it "really" means.

2. *Reducibility; emergence:* If we choose to see a hierarchical structure in Nature—if societies are composed of individuals, individuals of cells, and cells in their turn of molecules, then it makes sense to ask whether we may not "interpret" sociology in terms of the biology of individuals or "reduce" biology to physics and chemistry. This is a living methodological problem, but it does not seem to have been satisfactorily resolved. At first sight the ambition embodied in the idea of *reducibility* seems hopeless of achievement. Each tier of the natural hierarchy makes use of notions peculiar to itself. The ideas of democracy, credit, crime or political constitution are no part of biology, nor shall we expect to find in physics the concepts of memory, infection, sexuality, or fear. No sensible usage can bring the foreign exchange deficit into the biology syllabus, already grievously overcrowded, or nest-building into the syllabus of physics. In each plane or tier of the hierarchy new notions or ideas seem to emerge that are inexplicable in the language or with the conceptual resources of the tier below. But if in fact we cannot "interpret" sociology in terms of biology or biology in terms of physics, how is it that so many of the triumphs of modern science seem to be founded upon a repudiation of the doctrine of irreducibility? There is a problem here

to which methodologists can and do make valuable and illuminating contributions.[8]

[8] See, for example, E. Nagel, *The Structure of Science* (New York, 1961); A. Pap, *An Introduction to the Philosophy of Science* (London, 1963). The problem of "reducing" sociology to biology goes back at least to John Stuart Mill: "The laws of the phenomena of society are, and can be, nothing but the laws of the actions and passions of human beings united together in the social state . . . Human Beings in society have no properties but those which are derived from, and may be resolved into, the laws of the nature of individual men" (*System of Logic* [7th ed., London, 1868] Book VI: chap. VII, § 1).

The examples given in the text were chosen to make the point that many ideas belonging to a sociological level of discourse make no sense in biology and that many biological ideas make no sense in physics, but it is important not to forget that this restriction on the flow of thought works one way only. Nothing disqualifies the inclusion of physical or chemical propositions in the biological or social sciences. That gold should be used as a currency standard depends in part on its being uncorruptible by rust and the ravages of lepidoptera, and to "explain" this property we must investigate its physico-chemical properties or seek guidance from those who do so. The explanation of why an unexpectedly high proportion of Nigerians enjoy an inborn resistance to subtertian malaria turns on a specially detailed knowledge of the structure of the hemoglobin molecule. Examples of this kind are limitless. There is a sense in which the social sciences comprehend biology and make use of biological notions, and in which sociology is empirically and conceptually the richer subject. The same could be said of the biological sciences *vis-à-vis* physics and chemistry. Yet there is also a sense in which physics comprehends biology, and biology in its turn the social sciences, as the more general sciences comprehend the more particular.

These are exasperatingly vague statements, and they sound paradoxical: how can sociology be empirically and conceptually richer than biology if there is a sense in which biology comprehends it?

The only way I can make the case is by appealing to an analogy between the hierarchy of the empirical sciences and the hierarchy of classical geometries as it was envisaged by "the greatest synthesist that geometry has ever known," Felix Klein. The reader must judge for himself whether or not the analogy is illuminating. In what follows I shall not strive after a rigor I am unqualified to achieve. (See J. L. Coolidge, *A History of geometrical Methods* [Oxford, 1940]; E. T. Bell, *The Development of Mathematics* [New York, 1940]. Klein's great synthesis

3. *Causality:* The problems raised by the notion of necessary connexion, and the discussion of its actual and proper use. No one who has studied the slovenly and sometimes actively misleading ways in which genet-

---

was adumbrated in 1872, and the best account of it is his own: *Elementary Mathematics from an Advanced Standpoint: Geometry,* trans. E. R. Hedrick and C. Noble [London, 1939; 1st German edition, 1908]. By "Classical" geometries I mean those in which geometric objects are represented as being contained (as by a vessel) in a space that is independent of them.)

In Klein's conception, a geometry is the invariant theory of a certain specified group of geometric operations, i.e. it is the class of statements describing the properties of geometric objects that remain *un*changed under the transformations to which they are subjected. A transformation may be thought of as a substitution of one set of points for another, keeping the same coordinate system; or, alternatively, as a substitution of one coordinate system for another, the points themselves being thought to remain unchanged. Whichever way we choose to think of it, the transformation can be described in a geometric language—we can speak of displacements, rotations, inversions, etc.—or they may be defined by analytic (algebraic) formulae, "mapping functions," which are rules for exchanging the new points (or new coordinates) for the old. (A "group" of transformations is a set of which each member has an inverse, and which is such that the successive performance of any two transformations is itself a member of that set. "Displacements" in space form a group, because any displacement [say from position *A* to *B*] has its inverse [*B* to *A*] which restores the *status quo,* and the product of two successive displacements is itself a displacement. In chess, pawns' moves have no inverse; knights' moves have an inverse, but the product of two knights' moves is not a knight's move. Neither forms a "group".)

In this scheme of codification, metric, Euclidian and affine geometries and topology may be said to form a hierarchy: we can pass from one to the other by progressively relaxing the conditions imposed by the rules of transformation, or (in the other direction) by making them progressively stricter. Metric geometry is the most highly restricted: the group of operations that defines it consists only of translations, rotations and inversions. The invariant theory of this group of operations is the richest in geometric concepts: it will contain a superabundance of theorems to do with isosceles triangles, regular polygons and with degrees of curvature and angularity; it can make use of the idea of scalar

icists were at one time wont to discuss the relationship between "gene" and "character" will dismiss the problem as dead or unworthy of attention. To bring the point home, let me make four successive statements about the

---

distance also, for the distance between two points is invariant under the transformations of the metric group—transformations which conserve all properties associated with size and shape.

The Euclidian group of transformations is a little more permissive: symmetrical magnification is allowed and the concepts of size and metric distance therefore disappear, though the notions of (for example) square and circle are retained, and indeed all properties to do with *shape,* which is invariant.

Affine geometry is specified by a group of transformations which (in geometrical terms) allows for uniform magnification, but to different degrees in the three dimensions of space. The concepts of square and circle and size of angle are now meaningless, since the properties that define them are not invariant under transformation, but linearity and parallelism remain, and theorems to do with ellipses and parallelograms. Geometric objects which are Euclidian transforms of each other are described as "similar"; when they are affine transforms of each other they are sometimes described as *homeographic.* A mechanical or architectural drawing is a special kind of homeograph of the object it represents. (Projective geometry, familiar from perspective drawing, includes affine geometry as a special case, but it does not belong in the hierarchy under discussion because the transformations that define it allow points to be "carried to infinity." In projective geometry, linearity remains, but not parallelism; and circle, ellipse, etc., give way to the more general notion of a conic section.)

Topology is the most permissive of the four geometries, for nothing is required of the transformations that define it except that they should be continuous and should bring the transformed points into a one-to-one correspondence with the points they replace. A topological transformation may be represented geometrically by an arbitrary plastic deformation, such as a geometric figure would undergo if it were drawn upon a sheet of rubber which was thereupon stretched or twisted in any way that did not tear it. Figures related to each other by continuous plastic deformation retain the kind of primitive likeness that is described as *homeomorphy.* Obviously all simple geometric notions have now lost their meaning, but certain very elementary properties remain, e.g., the order of points on a line, relationships of insideness and outsideness of

role of the Y chromosome in the determination of sex in man. In man

(a) the possession of a Y chromosome is the cause of maleness;

---

closed figures, the "sidedness" of surfaces (in these enlightened days all children are taught to play with Möbius strips).

It follows from the way in which they were derived that all the theorems of topology are "true in" (and all topological concepts make sense in) affine geometry, that the theorems and concepts of affine geometry are part of Euclidian geometry, and so on. As we pass down the series, topology—affine geometry—Euclidian geometry—metric geometry, we may note that: (a) each geometry is a special case of its predecessor, i.e. is derived by imposing special restrictions upon or defining a subgroup within the one preceding it; (b) all theorems of one geometry are also theorems in its successors; (c) new concepts (e.g. of parallelism, circularity or shape) "emerge" at each level which have no meaning and cannot be envisaged at an earlier level; and (d) there is a progressive enrichment in the number and variety of concepts and the particularity and degree of detail of the theorems.

My argument is that much the same kind of relationship holds between the elements of a hierarchy of the empirical sciences, taken in the order physics—chemistry—biology—sociology. The hierarchy is in this case compositional (inasmuch as men are made of molecules and societies of men), the rules of transformation are causal rather than algebraic, and the theorems are of empirical origin, but (*mutatis mutandis*) the relationships *a,b,c,d* are valid nevertheless. In the light of this interpretation the ideas of *reducibility* and *emergence* are no longer mystifying; nor is it self-contradictory to say that sociology comprehends biology even though it is, formally speaking, a "special case" (in the sense that "societies" are only a subclass of all possible systems of interaction between individuals, just as living organisms represent a subclass of all possible configurations or systems of interaction between molecules). It is a sociological truth as well as a physical truth that the atomic weight of sulphur is 32. The trouble about such a statement is not that it is false or meaningless in the social sciences, but that it is unimportant or dull. The same, however, would apply to a topological theorem in the context of metric geometry, e.g. "every right-angled isosceles triangle divides a plane into a part inside it and a part outside it." "What of it?" is the natural comment in either case.

(*b*) the possession of a *Y* chromosome causes the difference between male and female characteristics;

(*c*) the substitution of a *Y* chromosome for one of the two *X* chromosomes causes the difference between male and female characteristics;

(*d*) there is a wide but definable class of genetic and environmental situations in which the substitution of a *Y* chromosome for one of the two *X* chromosomes causes the difference between male and female characteristics.

These statements mark four stages in the refinement of a vague and barely articulate but obviously "significant" idea. The first is scientifically semi-literate; the fourth, though long and clumsy, is pretty well acceptable. The notion of causality pervades the whole of science, and no one science has any special claim to adjudicate upon its usage.[9] The existence of problems of this kind is justification enough for the existence of a

[9] The terminology by which we speak of a gene substitution's causing a character difference was quite largely influenced by L. T. Hogben, *Nature and Nurture* (London, 1933). Many think it clumsy and advocate more elaborate formulations (e.g. J. H. Woodger, *Biology and Language* [Cambridge, 1952]). However, the usage is one that comes naturally to experimental scientists. When we carry out an experiment of ordinary unifactorial design (one factor or circumstance varied, the others kept constant), the result of the experiment is the *difference* between two sets of readings (or two sets of phenomena or two events), namely those recorded in the experiment itself and those recorded in its 'control'; and the inference we are entitled to draw is that the difference between the starting conditions was the cause of the difference between the two sets of results. This is precisely the genetic usage. In everyday life, of course, we speak of the causes of events, phenomena, or states of affairs, but the cause we have in mind, when analyzed, usually turns out to be the cause of a difference between *what was* and *what might have been;* between what did happen and what might have happened if the antecedents had themselves been different.

science or area of discourse known as scientific methodology, even if its task falls short of expounding the nature of scientific method as a whole. I must not therefore be thought to imply that the pursuit of methodology is a waste of time.

I called this lecture "The Problem Stated," and the
problem is twofold: that which is embodied in the question "What is the scientific method?" and that which is
embodied in the fact that scientists pay no serious attention to the answer. But answers have been given, in
spite of the scientists' indifference, and in my next lecture I shall ask whether the doctrine of *induction* provides a good enough approximation to the truth.

II. MAINLY ABOUT INDUCTION

1

TOWARDS THE END of my first lecture I said that scientific methodology had important but limited tasks to perform in the analysis or clarification of certain ideas that were common to all the sciences. I mentioned three of them: the ideas of validation or justification, of reducibility and emergence, and of causality. The older methodologists would not have been satisfied with such limited ambitions. Their intention was to lay bare the whole structure of scientific reasoning, to expound all the distinctive acts of thought that enter into scientific discovery and the enlargement of the understanding.

For more than a hundred years the English-speaking world has been dominated by the opinion that scientific reasoning is of a special kind, *inductive:* an opinion so strongly advocated by such skillful and persuasive thinkers that even when many of the principles of induction have been repudiated or allowed to fade away we still remain in an unconscious bondage to a number of inductive practices and habits of thought.

To the question "What is Induction?" there is no simple answer, even when we eliminate mathematical induction (a special usage) and that humble form of induction which assures us that what is true of each must be true of all. Inductivism is a formulary of beliefs, a complex of attitudes and practices having to do with the nature of science and of scientific enquiry, and in what follows I shall do my best to give a fair account of the various elements that enter into it. Although much of what I shall say will be critical, I do not want to give the impression that, in my opinion, induction has no place in science at all. There are indeed certain limited and special occasions on which we carry out induction according to the rules of Bacon and of Mill, but I shall defer mention of them until later on.

One point should be made clear from the beginning. In the traditional view of induction, that which is embodied in dictionary definitions, induction is "arguing from the particular to the general" where deduction is arguing from the general to the particular. Induction, then, is a scheme or formulary of reasoning which somehow empowers us to pass from statements expressing particular "facts" to general statements which comprehend them. These general statements (or laws or principles) must do more than merely summarize the information contained in the simple and particular statements out of which they were compounded: they must add something, say more than that which has been said already—for what would be the use of a "Law of Nature" which merely authenticated or conferred respectability upon the phenomena already known to obey it? Inductive reasoning is *ampliative* in nature. It expands our knowledge, or at all events our pretensions to knowledge.

This is all very well, but the point to be made clear

is that induction, so conceived, cannot be a logically rigorous process. It cannot (as deduction can, if properly executed) lead us with certainty to the truth. Mill believed it could do so, but John Venn and C. S. Peirce and others flatly disagreed with him,[10] and it is their opinion that has prevailed. I shall waste no time attacking a position that is no longer defended. No process of reasoning whatsoever can, with logical certainty, enlarge the empirical content of the statements out of which it issues. If it could indeed do so then all scientific research could be carried out in a recumbent posture, with the eyes half closed.

2

Now let me discuss one by one the shortcomings of the inductive style of reasoning, as I see it; for in finding fault with induction I am by implication helping to define the properties which a really adequate methodology should enjoy.

§ 1. At the very heart of induction lies this innocent-sounding belief: that the thought which leads to scientific discovery or to the propounding of a new scientific theory is logically accountable and can be logically spelled out. Even if they are not apparent at the time (because they have been short-circuited or speeded up), a retrospective analysis can reveal the processes of reasoning and the logically motivated actions which conduct the scientist towards what he be-

[10] Amongst the others were Dugald Stewart (writing fifty years before Mill) and Stanley Jevons. On this point, see my *The Art of the Soluble* (London, 1967), pp. 136-137.

lieves to be the truth. There is a grammar of science, and the language of science can be parsed.

This is quite different from saying that *given* some belief or opinion or would-be natural law, no matter what its origin (whether by research, by revelation, or in a dream), *then* its acceptability can be tested by procedures that involve the use of logic. In the inductive view, it is the process of *getting* an idea or formulating a general proposition that can be logically reasoned out. It follows that, in the inductive scheme, discovery and justification form an integral act of thought. That which leads us to form an opinion is also that which justifies our holding the opinion. The intellectual processes that conduct us towards a generalization are themselves the grounds for supposing it to be true.[11]

This concept of the inductive process must have arisen out of a misleading formal analogy with *de-*duction. In deductive reasoning, e.g. in Euclid, we discover or uncover a theorem by reasoning which, if we have carried it out correctly, guarantees the theorem to be true if the axioms or premisses are true. Our ability

[11] See Whewell's critique of Mill in chap. 22, particularly pp. 262 onwards, of *The Philosophy of Discovery* (London, 1860). The later editions of Mill's *System* challenge Whewell's objections (Book III, chap. VI, § 6), though not very convincingly. Mill ends rather lamely by saying, with reference to his Four Methods:

"If discoveries are ever made by observation and experiment without Deduction, the four methods are methods of discovery: but even if they were not methods of discovery, it would be not less true that they are the sole methods of Proof...."

I feel that Mill was handicapped by his deeply held conviction that induction and deduction were processes of the same stature or logical authority, and in the next paragraph of the main text I try to show how misleading this belief can be.

to deduce Pythagoras' Theorem from Euclid's axioms—
i.e. to discover Pythagoras' Theorem in Euclid's axioms
—is in itself our justification for believing it to be valid.
In a purely formal sense, therefore, discovery and justification are the same process in deductive logic; but
the qualification "in a purely formal sense" is very
important. It is most unlikely that more than a tiny
minority of mathematical theorems were ever in fact
arrived at, "discovered," merely by the exercise of deductive reasoning. Most of them entered the mind by
processes of the kind vaguely called "intuitive"; deduction or logical derivation came later, to justify or falsify
what was in the first place an "inspiration" or an intuitive belief. This is seldom apparent from mathematical
writings, because mathematicians take pains to ensure
that it should not be. Deductivism in mathematical
literature and inductivism in scientific papers are simply
the postures we choose to be seen in when the curtain
goes up and the public sees us. The theatrical illusion is
shattered if we ask what goes on behind the scenes. In
real life discovery and justification are almost always
different processes, and a sound methodology must
make it clear that they are so.

§ 2. Inductive theory insists on the primacy of Facts:
of propositions that put on record the simple and un-
complicated evidence of the senses. Karl Pearson [12]
was a great believer in facts:

> The classification of facts and the formation of absolute judg-
> ments upon the basis of this classification . . . essentially sum
> up the *aim and method of modern science.*

[12] *The Grammar of Science* (3rd ed., London, 1911; 1st ed., 1892).
The passages I quote are from pp. 6 and 9 (see also p. 37). The italics
are Pearson's.

The classification of facts, the recognition of their sequence and relative significance, is the function of science.

Pearson felt that the study of facts was conducive not only to good science but to right-mindedness in general:

Modern Science, as training the mind to an exact and impartial analysis of facts, is an education specially fitted to promote sound citizenship.

It may therefore seem downright subversive to question the primitive authenticity of facts or to cast doubt upon evidence in the form in which it is delivered to us by the senses; it is worse still to ask, as Whewell did, how often "facts" can be stripped of a mask of interpretation and theory. It is very un-English, to be sure, for to put such a question is to challenge the greatest philosophic tradition of the English-speaking world, the tradition of philosophic empiricism which we inherit from John Locke. Nothing enters the mind except by way of the senses (its fundamental principle goes); and though the senses may sometimes be clouded, though we may sometimes be the victims of deception and illusion, yet if we can only get at it in its primitive simplicity, the evidence of the senses is the foundation of all knowledge that is not merely a repartition of ideas or words. There is an *essential* trustworthiness about the evidence of the senses, and therefore about the simple observational statements which put that evidence on record.

It won't do, of course. No one now seriously believes that the mind is a clean slate upon which the senses inscribe their record of the world around us: that we take delivery of the evidence of the senses as we take delivery of the post. "Everything that reaches consciousness is utterly and completely adjusted, simpli-

fied, schematized, interpreted," said Nietzsche,[13] in one
of his exhilarating outbursts of common sense. Innocent,
unbiased observation is a myth: "experience is itself a
species of knowledge which involves understanding,"
said Kant.[14] What we take to be the evidence of the
senses must itself be the subject of critical scrutiny.
Even the fundamental principle of empiricism is open
to question, for not all knowledge can be traced back
to an origin in the senses. We inherit some kinds of
information. A bird's song is in some sense the tran-
scription of a chromosomal tape recording, and the same
goes for the entire repertoire of all that can properly
be called "instinctual" behavior.[15]

[13] Friedrich Nietzsche, *The Will to Power,* trans. A. M. Ludovici
(London, 1910). See especially Vol. 2, §§ 466-617. Nietzsche's ideas on
methodology are worth serious attention; those who have dismissed
him as the author of *Thus Spake Zarathustra* will be surprised at the
strength and insight of his opinions on methodology and the theory of
knowledge.

[14] *The Critique of Pure Reason,* trans. N. Kemp Smith (2nd ed.,
London, 1933), p. 22.

[15] To my mind the most striking examples of "genetically pro-
grammed" behavior sequences are not those in which an animal reared
in isolation from birth or hatching turns out to "know" how to sing a
particular song or build a nest, but those in which a suitable stimulus
transforms one behavior pattern into another. A wealth of examples
relating to sexual behavior will be found in D. S. Lehrman, pp. 1268-1382
of *Sex and Internal Secretion* (3rd ed., Baltimore, 1961). Thus (p.
1299):

"Domestic cocks take no part in the care of the young, and sometimes
even kill chicks that are confined with them. [Cocks cannot be induced
to incubate eggs by prolactin injections, but] a number of workers have
reported that prolactin induces cocks to cluck, to lead, and to protect
chicks under their body and wings."

As to the level of detail of the genetic program, we may note (p. 1351),
that

"Courting male canaries sometimes dangle a piece of string or cotton

§ 3. Although inductive exercises often begin with an injunction to assemble all the "relevant" information (relevant to what?), inductive theory provides no *formal* incentive for making one observation rather than another. Why indeed do we not count and classify the pebbles in a gravel pit (see note 6)? This is a subject on which Coleridge had a number of pointed and sensible things to say.[16] Any adequate account of scientific method must include a theory of incentive or special motive; must contain a canon to restrict observation to something less than the whole universe of observables. We cannot browse over the field of nature like cows at pasture.

§ 4. In my first lecture I said (with induction in mind) that scientific methodology was almost obsessively preoccupied with problems of justification and ascertainment—with laying down the conditions under which the views we hold should be judged right or wrong, or with quantifying the degree of confidence we should have in them. When scientific research is studied

---

before the female. Shoemaker . . . found that female canaries injected with testosterone propionate postured like courting males, and also engaged in this string carrying behavior, reminiscent of the carrying of nesting material."

The argument that all instinctual behavior must at some time have been learned, though not necessarily in the generation in which it is acted out, will not hold water, for a chromosomal aberration might conceivably produce a purely endogenous change of behavior pattern. For an early discussion of "inherited ideas," see chap. IV. in E. Mach, *The Analysis of Sensations,* trans. C. M. Williams and S. Waterlow (London, 1914; 1st ed. 1885).

[16] Coleridge *On Method:* above, note 3. See also *The Art of the Soluble,* pp. 143-144.

on the hoof, so to speak, we find that very few theories
are utterly discredited in the style in which (for example) Thomas Henry Huxley demolished Goethe's
and Oken's Vertebral Theory of the skull.[17] Theories
are repaired more often than they are refuted, and a
methodology of rectification (a logical variant of negative feedback) is something we shall expect to find in
any satisfactory formal account of scientific reasoning.
Sometimes theories merely fade away: no one now
believes in the doctrine of "Protoplasm," but no one
to my knowledge has ever refuted it.[18] More often they
are merely assimilated into wider theories in which
they rank as special cases. The law of recapitulation
and the germ-layer theory have not been shown to
be "wrong." They have simply lost their identity and
their special significance in an improved understanding
of the mechanism of development. They have been
trivialized.[19]

[17] In his famous Croonian Lecture to the Royal Society: *Proc. Roy.
Soc.,* ser. B, **9** (1857-1859): pp. 381-457 (reprinted in *The Scientific
Memoirs of T. H. Huxley,* eds. M. Foster and E. Ray Lankester
(London, 1898) **1**: pp. 538-606.

[18] *The Art of the Soluble,* pp. 104-106.

[19] *Recapitulation.* If a human being has embryonic gill pouches before
his lungs begin to form, that is not because development is obliged to
rehearse evolution, but rather because the lung *is* a special sort of gill
derivative, just as jaws are gill arches of a special kind, or limbs a special
sort of fin. It is therefore understandable from the point of view of developmental mechanics that the embryos of higher vertebrates should
sometimes recapitulate the *embryonic* history of their ancestors (see
G. R. de Beer, *Embryos and Ancestors* [Oxford, 1940]). The Law of
Recapitulation is to some extent true of what might be called "serial"
evolution, the substitution of one terminal developmental episode for
another, as in the examples cited above; but there is no reason why it
should be expected to be true of evolutionary processes that foreshorten

§ 5. One of the most damaging charges against inductivism was brought for the first time, I believe, by Lord Macaulay,[20] though not in the form in which I shall present it here. Inductivism gives no adequate account of scientific fallibility, fails altogether to explain how it comes about that the very same processes of thought which lead us towards the truth lead us so very much more often into error.

Methodologists who have no personal experience of scientific research have been gravely handicapped by their failure to realize that nearly all scientific research leads nowhere—or, if it does lead somewhere, then not in the direction it started off with. In retrospect we tend to forget the errors, so that "The Scientific

the life history (neoteny) or which introduce novelties during embryonic or juvenile stages of development.

Germ Layer Theory. The embryos of many vertebrates have a layered or laminate structure, and there is a surprising degree of regularity in the nature of the adult organs into which the several layers develop—outermost layer into skin and nervous system, innermost layer into the viscera, and so on. The layered structure, as we see it under the microscope, now merely reminds us of something unknown to the embryologists who propounded the germ layer theory, viz. that the morphogenetic exercises of early vertebrate development consist of the sliding, folding, stretching and glassblower-like manipulations of cellular sheets as they come to take up the remarkably uniform stations characteristic of the early axiate embryo of vertebrates.

The Law of Recapitulation and the Germ Layer Theory are not "wrong": they can be used to make a point about development, even if the point is now hardly worth making. We call them to mind only to help us understand the newer reasoning by which they have been superseded, for something akin to a principle of recapitulation works here too, in the history of ideas.

[20] In his extended review (often reprinted in collections of Macaulay's "Essays") of Montagu's edition of Bacon's *Works: Edinburgh Review,* July, 1837.

Method" appears very much more powerful than it really is, particularly when it is presented to the public in the terminology of breakthroughs, and to fellow scientists with the studied hypocrisy expected of a contribution to a learned journal. I reckon that for all the use it has been to science about four-fifths of my time has been wasted, and I believe this to be the common lot of people who are not merely playing follow-my-leader in research.

Why do scientists hold or come to formulate erroneous opinions? That, surely, is a central problem of methodology. For traditional inductive theory, as Popper has explained,[21] error must be held to arise from misapprehension of the facts, the data; from a misreading of that Book of Nature in which the truth resides if only we had the skill and clarity of vision to discern it. If the facts are misapprehended through blindness or prejudice, then the inferences logically induced from them must be mistaken; and so we are led into error. (This follows the deductive analogy pretty closely, for if in deduction the theorems are empirically wrong it must be because the axioms were empirically wrong, unless there has been an avoidable mistake of reasoning.) If only we had a clear and perspicacious vision the truth would make itself apparent to us, as it was to the chap in Voltaire who apprehended all human knowledge in a matter of weeks because, until he grew out of savage innocence, his mind had been preserved from the prejudices and ideological corruptions of prior learning.[22]

[21] K. R. Popper, *Conjectures and Refutations* (London, 1963), pp. 3-30.

[22] I came across Voltaire's *L'Ingénu* through J. Agassi's sparkling

This cannot be the whole truth about error. What shows a theory to be inadequate or mistaken is not, as a rule, the discovery of a mistake in the information that led us to propound it; more often it is the contradictory evidence of a new observation which we were led to make *because* we held that theory. Error or insufficiency is shown up by a critical process applied in retrospect; only seldom is it due to a failure of apprehension, to a dullness or cloudiness of vision. It is a bad mark against inductivism that it provides us with no acceptable theory of the origin and prevalence of error.

§ 6. What are we to make of *luck* in our methodology of science? In the inductive view, luck strikes me as completely inexplicable; it can arise only from the gratuitous obtrusion of something utterly unexpected upon the senses; it is like winning a prize in a lottery in which we did not buy a ticket. To buy a ticket is to define a category of expectations, and then the reason why we won is obvious: we were in luck; for once in a way our hopes were gratified. We have Fontenelle's and Pasteur's word for it that luck makes sense only against a background of prior expectations. Ever since his experiences in the First World War, Alexander Fleming had been deeply concerned by the problem of infected

essay *Towards an Historiography of Science* (History and Theory **2**, s'Gravenhage, 1963). Brought up in feral innocence, the hero "faisait des progès rapides dans les sciences . . . La cause du développement rapide de son esprit était due à son éducation sauvage presque autant qu'à la trempe de son âme: car, n'ayant rien appris dans son enfance, il n'avait point appris de préjugés. Son entendement, n'ayant point été courbé par l'erreur, était demeuré dans toute sa rectitude. Il voyait les choses comme elles sont . . ." (chap. **XIV**). *L'Ingénu* was Inductive Man.

wounds. It was his lifelong ambition to discover a non-toxic antibacterial agent, and in penicillin he found a winner—by luck, if you like; but he held a ticket which entitled him to win a prize.[23]

§ 7. In § 5 I said that the shortcomings of scientific theories were usually revealed by the exercise of a critical process. Classical inductive theory reveals no clear grasp of the *critical* function of experimentation. "Experiments" are of several different kinds—in a moment I shall mention four—and what I shall describe as inductive or "Baconian" experimentation is only one of them. Let me first explain what I conceive Baconian experimentation to be.

If we believe that the initiative for scientific research lies in observation, that scientific knowledge grows out of the evidence of the senses, then our first duty as scientists must be to observe nature faithfully, intently, and without misleading preconceptions. Let us then imagine ourselves wandering through a sylvan or pastoral world and recording our observations. Obviously we could not observe enough to sustain the growth of science. We could spend a lifetime observing nature without once seeing two sticks rub together, seeing sunlight refracted through a crystal, or witnessing the distillation of fermented liquor. Francis Bacon therefore charged us to contrive or invent experiences; to mess about, we might say today (the phrase would not have come naturally to Bacon); to indulge in ex-

---

[23] There was indeed an element of blind luck in the discovery of penicillin. See my review of J. D. Watson's *The Double Helix* in the *New York Review of Books* (March 28, 1968).

periential play. Let us try experiments [24] with the
burning glass and lodestone and rubbed amber; let us
distill liquors not once but twice successively, to see
what happens. Nature is an actress with a prodigious
repertoire: give her an opportunity to perform.

It was these contrived experiences, invented happen-
ings, that Bacon called *experiments,* and this is still
the vernacular usage of today.[25] Experimentation in
Bacon's sense is not a critical procedure. Its purpose
is to nourish the senses, to enrich the repertoire of
factual information out of which inductions are to be
compounded. It is that enlargement of experience
which in the inductive view, cannot but lead to an
enlargement of the understanding.

Experiments are of at least four kinds:

(i) Inductive or Baconian experiments such as those
I have just described ("I wonder what would happen
if . . .").

(ii) Deductive or Kantian experiments, in which we
examine the consequences of varying the axioms or pre-
suppositions of a scheme of deductive reasoning ("let's
see what happens if we take a different view"). In
introducing the greatest intellectual exploit in the history of philosophy, Kant invites us, "by way of an experiment in pure reason," to abandon the commonplace

---

[24] Note the terminology: nowadays we *do* or *carry out* experiments;
we no longer *make* or *try* experiments.

[25] Mill's usage is essentially the same, though it is naturally more
sophisticated: see his *System,* Book III, chap. VII, §§ 3,4. "Artificial"
experimentation is an extension of "pure" observation: "it enables us to
obtain innumerable combinations of circumstances which are not to be
found in nature, and so add to nature's experiments a multitude of
experiments of our own."

view that our sensory intuitions conform to the constitution of objects, and to examine the consequences of supposing that objects conform to the constitution of our faculty of intuition. "This experiment succeeds as well as could be desired"; it explains how we may have knowledge that is independent of all experience, and leads Kant to propose that space and time are not objects of but *forms* of sensory intuition.[26]

Perhaps the best examples of "pure" Kantian experiments are those which led to the formulation of the classical non-Euclidian geometries (hyperbolic, elliptic) by Saccheri, Lobachevsky, Bolyai, Riemann, and others.[27] It is ironical that this should be so, for as W. K. Clifford cleverly discerned,[28] the non-Euclidian geome-

[26] Kant, *op. cit.,* note 14. To call such experiments "Kantian" is by no means to imply that Kant did not have a clear understanding of the critical functions of experimentation; Kant's grasp of scientific methodology was remarkable (see also note 39 below).

[27] We may construe the classical non-Euclidian geometries as being derivable by the experiment of replacing Euclid's axiom of parallels (or its formal equivalents) by the "obtuse angle hypothesis" or the "acute angle hypothesis" (or their formal equivalents), so generating geometries representable in spaces of constant positive curvature or constant negative curvature respectively (see W. K. Clifford, note 28 below). In this scheme Euclidian geometry is a special case, representable in a space of zero curvature. It has often been pointed out that the formulation of non-Euclidian geometries had a profound influence on our conception of mathematical reasoning and the nature of the "truths" to which it leads.

[28] See in particular W. K. Clifford's four lectures on "The Philosophy of the Pure Sciences" (*Lectures and Essays,* ed. Leslie Stephen and F. Pollock [London, 1879] 1: pp. 254-340); see also his lecture "On the Aims and Instruments of Scientific Thought," *op. cit.,* pp. 124-157. Clifford had a deep understanding of mathematical thought and his lectures are a delight to read, but his views on the empirical sciences are not always convincing. His criticism of Kant must be reappraised in the

tries might be thought to call into question the very truths concerning *a priori* knowledge to which Kant supposed that his own audacious experiment had led him.

(iii) Critical or Galilean experiments: actions carried out to test a hypothesis or preconceived opinion by examining the logical consequences of holding it. Galilean experiments discriminate between possibilities. I shall say more about them in Lecture III.

(iv) Demonstrative or Aristotelian experiments, intended to illustrate a preconceived truth and convince people of its validity. In *Plus ultra* (1668), a puff for the New Science, Joseph Glanvill wrote of Aristotle that "he did not use and imploy Experiments for the erecting of his Theories: but having arbitrarily pitch'd his Theories, his manner was to force Experience to suffragate, and yield countenance to his precarious Propositions" (p. 112). Thomas Sprat took a very poor view of experimentation in this style—". . . a most venomous thing in the making of sciences; for whoever has fix'd on his Cause, before he has experimented, can hardly avoid fitting his Experiment to his own Cause . . . rather than the Cause to the truth of the Experiment it self" (*History of the Royal Society* [1667], p. 108).

Most original research begins with Baconian experimentation. We undertake to study a problem or a phenomenon for many different reasons: because it is interesting or important, because we have been led to it

---

light of our present knowledge that Kant was "in on" the earliest stages of speculation about the necessary validity of Euclidian geometry: see A. C. Ewing (*A Short Commentary on Kant's Critique of Pure Reason* [London, 1938]).

through earlier researches, or because we have been asked or told to do so; but whatever our motives may have been, the first thing to do is to find out what actually happens, what is in fact the case ("let's find out in a bit more detail what it is we are actually studying"). We describe and annotate the phenomena when they are made to take place under certain well-defined and well-regulated conditions. In the meanwhile we begin to form opinions about the nature of the causal mechanisms at work and the relationship of the phenomena to others, and only critical experimentation can discriminate between them. Sciences which remain at a Baconian level of development, as (for example) the study of animal behavior did until its modern renaissance,[29] amount to little more than academic play.

I said near the beginning of this lecture that there were certain situations in which we undoubtedly use inductive reasoning—and in very much the style envisaged by Bacon or, more explicitly, by Mill. Suppose we have set up some "tissue cultures" of living cells using a variety of media which have subsequently been thrown away. Some of the cultures, but not all, have been ruined by bacterial infection, and we naturally wish to find out why. Mill's Five Canons [30] will conduct us towards the answer; or, if they do not, nothing will. Media common to all the cultures cannot have been responsible for introducing the infection. If the infected cultures, and they alone, were set up with a medium from a certain special source, then that medium was almost certainly responsible; and we shall be con-

[29] *The Art of the Soluble,* pp. 108-110.
[30] *System,* Book III, chaps. VIII and IX.

firmed in this interpretation if we find that the more heavily contaminated cultures were those in which a larger quantity of the medium under suspicion had been used. We are taken aback when a fuller study of the records shows that a number of cultures escaped infection although the supposedly infected medium had been used to prepare them, but it turns out that these anomalous cultures differed from those which were overtly contaminated by the use of a bactericidal ingredient which kept the infection down. And so on: the situation can be made as complicated as we please, but the reasoning which resolves it is straightforward and quite commonplace. It is "logical" in the sense that it can be carried out by formula or by rote, and it is, or can be conclusive if the empirical facts, as stated, represent the whole truth and nothing but the truth. Conversely, if our conclusions are wrong it must be because there was a mistake in the "facts" from which the induction started.[31]

[31] Mill's own examples of induction at work are strange reading today (*loc. cit.,* note 30), particularly when he chooses biological examples, as Whewell pointed out (*Philosophy of Discovery,* chap. XXII, § 41). Mill's methods, he adds, have a great resemblance to Bacon's, and we may note that Macaulay's scathing example of inductive reasoning (note 20) antedates both Whewell and Mill. It is no different in principle from my infected tissue cultures, but here the infection is gastrointestinal (if indeed an allergy is not to blame) :

"A plain man finds his stomach out of order. He never heard Lord Bacon's name. But he proceeds in the strictest conformity with the rules laid down in the second book of the *Novum Organum,* and satisfies himself that minced pies have done the mischief. 'I ate minced pies on Monday and Wednesday, and I was kept awake by indigestion all night.' This is the *comparentia ad intellectum instantiarum convenientum.* 'I did not eat any on Tuesday and Friday, and I was quite well.' This is the *comparentia instantiarum in proximo quæ natura data privantur.*

We turn naturally to inductive reasoning when we undertake a retrospective causal analysis of a state of affairs that is already "given." Was it not perhaps the goal of inductivism that all scientific reasoning should aspire to such a condition? Let us first assemble the data; let us by observation and by making experiments compile the true record of the state of Nature, taking care that our vision is not corrupted by preconceived ideas; *then* inductive reasoning can go to work and reveal laws and principles and necessary connections. A travesty of inductivism, you may think; but with something taken away for rhetorical overstatement, I think it not far short of the truth.

Let me now bring together the main points I have tried to make in this second lecture. Inductivism is a complex of opinions about the origin and character of scientific reasoning, and though it all hangs together in a creaking and groaning way it also has a number of ruinous shortcomings, the study of which (I said) should help us to plan the prospectus of a sounder methodology. In particular:

1. Inductivism confuses, and a sound methodology must distinguish the processes of discovery and of justification.

'I ate very sparingly of them on Sunday, and was very slightly indisposed in the evening. But on Christmas Day I almost dined on them, and was so ill that I was in great danger.' This is the *comparentia instantiarum secundum magis et minus.* 'It cannot have been the brandy which I took with them. For I have drunk brandy daily for years without being the worse for it.' This is the *rejectio naturarum.* Our invalid then proceeds to what is termed by Bacon the *Vindemiatio,* and pronounces that minced pies do not agree with him."

Here too is a retropsective causal analysis of "data," i.e. of information given. For "minced pies" read "mince pies" throughout.

2. The evidence of the senses does not enjoy a necessary or primitive authenticity. The idea, central to inductive theory, that scientific knowledge grows out of simple unbiased statements reporting the evidence of the senses is one that cannot be sustained.

3. A sound methodology must provide an adequate theory of special incentive—a reason for making one observation rather than another, a restrictive clause that limits observation to something smaller than the universe of observables.

4. Too much can be made of matters of validation. Scientific research is not a clamor of affirmation and denial. Theories and hypotheses are modified more often than they are discredited. A realistic methodology must be one that allows for repair as readily as for refutation.

5. A good methodology must, unlike inductivism, provide an adequate theory of the origin and prevalence of error . . .

6. . . . and it must also make room for luck.

7. Due weight must be given to experimentation as a critical procedure rather than as a device for generating information; to experimentation as a method of discriminating between possibilities.

Inductivism falls far short of being an adequate scientific methodology, and in my third lecture I shall study the credentials of an altogether different scheme of thought.

III. MAINLY ABOUT INTUITION

1

I SPENT MY second lecture finding fault with induction, excusing myself as I did so on the grounds that I was working my way towards the prospectus of a useful and realistic methodology. Of inductivism, I said that it was a complex formulary of methodological practices and beliefs of which the two most important elements were these:

(*a*) Observation is the generative act in scientific discovery. For all its aberrations, the evidence of the senses is essentially to be relied upon—provided we observe nature as a child does, without prejudices and preconceptions, but with that clear and candid vision which adults lose and scientists must strive to regain.

(*b*) Discovery and justification make one act of reasoning, not two. Inductive logic embodies both a rite of discovery and a ritual of proof. Scientific inference can always be made logically explicit, in retrospect if not at the time it was actually carried out.

So much of what I have said has been abstract that I feel I should now make amends by entertaining you with a couple of methodological caricatures.

Consider the act of clinical diagnosis. A patient comes

to his physician feeling wretched, and the physician sets out to discover what is wrong. In the inductive view the physician empties his mind of all prejudices and preconceptions and *observes* his patient intently. He records the patient's color, measures his pulse rate, tests his reflexes and inspects his tongue (an organ that seldom stands up to public scrutiny). He then proceeds to other, more sophisticated actions: the patient's urine will be tested; blood counts and blood cultures will be made; biopsies of liver and marrow are sent to the pathology department; tubing is inserted into all apertures and electrodes applied to all exposed surfaces. The factual evidence thus assembled can now be classified and "processed" according to the canons of induction. A diagnosis (e.g. "It was something he ate") will thereupon be arrived at by reasoning which, being logical, could in principle be entrusted to a computer, and the diagnosis will be the right one unless the raw factual information was either erroneous or incomplete.

Grossly exaggerated? Of course: as I said, a caricature; but, like a caricature, not exaggerated beyond all reason. It is obviously incomplete because no place has been found for flair and insight, and the enrichment that long experience brings to clinical skills. In Commencement Addresses and other uplifting declarations, clinicians who discourse upon the "spirit of medicine" will always point out that, while there is a large and profoundly important scientific element in the practice of medicine, there is also an indefinable artistry, an imaginative insight, and medicine (they will tell us) is born of a marriage between the two. But then (it seems to me) the speaker spoils everything by getting

the bride and groom confused. It is the unbiased observation, the apparatus, the ritual of fact-finding and
the inductive mumbo-jumbo that the clinician thinks
of as "scientific," and the other element, intuitive and
logically unscripted, which he thinks of as a creative art.

To see whether this apportionment of credit is a just
one, let us turn to another clinician in the act of diagnosis. The second clinician always observes his patient
with a purpose, with an idea in mind. From the moment
the patient enters he sets himself questions, prompted
by foreknowledge or by a sensory clue; and these questions direct his thought, guiding him towards new
observations which will tell him whether the provisional
views he is constantly forming are acceptable or unsound. Is he ill at all? Was it indeed something he ate?
An upper respiratory virus is going around: perhaps
this is relevant to the case? Has he at last done his
liver an irreparable disservice? Here there is a rapid
reciprocation between an imaginative and a critical
process, between imaginative conjecture and critical
evaluation. As it proceeds, a hypothesis will take shape
which affords a reasonable basis for treatment or for
further examination, though the clinician will not often
take it to be conclusive.[32]

This is a travesty too. The imagination cannot work
in vacuo: there must be something to be imaginative
about, a background of observation and Baconian experimentation, before the exploratory dialogue can

---

[32] Someone who goes to a clinic for a "check-up" will be exposed to a
battery of tests and observations and may well think himself the subject
of an inductive exercise. What is in fact happening is a sequential evaluation of likely hypotheses about what is or could be wrong with him.

begin. Nor have I explained the natural *progression* of thought that goes into clinical examination. But if I were asked which of the two accounts of the matter I thought the more helpful and realistic, I should without hesitation say the second. It is very far removed from induction, and belongs to an altogether different lineage of thought.[33] We find hints of it in Robert Hooke and Stephen Hales and Robert Boscovich. There are passages in Kant's lectures which reveal a clear understanding of what we have come to call the *hypothetico-deductive* method (I cite one of them below). By the middle of the nineteenth century it had established itself as the official alternative to the induction advocated by Mill. Whewell's *Philosophy of the Inductive Sciences* [34] (i.e. of the empirical sciences) is a masterpiece. Stanley Jevons (*The Principles of Science* [35]) can still be read with profit today; so also C. S. Peirce [36] and Claude Bernard.[1] The principal modern advocate and analyst of the theory is K. R. Popper; [37] my indebtedness to his writings will be obvious to anyone

[33] I have sketched the history of some of the main elements of the hypothetico-deductive system in my article "Hypothesis and Imagination" in *The Art of the Soluble,* pp. 131-155, and do not want to go over the same ground here.

[34] 1st ed., London 1840; 2nd ed., 1847. *The Philosophy of Discovery* (London, 1868) covers some of the same ground, but is not a substitute for the earlier work.

[35] 1st ed., London 1873; 2nd ed., revised 1877.

[36] *Collected Papers,* eds. C. Hartshorne and P. Weiss (Harvard, 1932-1935).

[37] *The Logic of Scientific Discovery,* a translation (with new appendices and footnotes) of *Logik der Forschung* (Vienna, 1934): *Conjectures and Refutations* (London, 1963). See also *The Critical Approach in Science and Philosophy,* ed. M. Bunge (London, 1964); M. Bunge, *Scientific Research* (2 vols., New York, 1967).

who knows them, but he must on no account be held
responsible for anything in what follows that may
seem unsound.

According to this second view, science, in its forward
motion is not *logically* propelled. Scientific reasoning is
an exploratory dialogue that can always be resolved into
two voices or two episodes of thought, imaginative
and critical, which alternate and interact. In the imaginative episode we form an opinion, take a view, make
an informed guess, which might explain the phenomena
under investigation. The generative act is the forma-
tion of a hypothesis: "we must entertain some hypothesis," said Peirce,[36] "or else forgo all further knowledge," for hypothetical reasoning "is the only kind of
argument which starts a new idea." The process by
which we come to formulate a hypothesis is not illogical
but non-logical, i.e. outside logic. But once we have
formed an opinion we can expose it to criticism, usually
by experimentation; this episode lies within and makes
use of logic, for it is an empirical testing of the logical
consequences of our beliefs. "If our hypothesis is sound,"
we say, "if we have taken the right view, then it follows
that . . ."—and we then take steps to find out whether
what follows logically is indeed the case. If our predictions are borne out (logical, not temporal predictions)
then we are justified in "extending a certain confidence
to the hypothesis" (Peirce again). If not, there must
be something wrong, perhaps so wrong as to oblige us
to abandon our hypothesis altogether.

A scientific theory can be thought of as a complex,

logically bonded polymer built out of the following elementary or monomeric form: [38]

LOGICAL SYNTAX	Hypotheses, axioms, postulates, premises, assumptions (etc.)	SEMANTICS
	Deductions, theorems, logical inferences, consequences, predictions (etc.)	

This elementary theory is supported by a *metatheory* which specifies the rules of deduction or statement-transformation ("logical syntax") and adjudicates upon the meanings of the empirical terms which it employs, i.e. says what they stand for ("semantics"). Postulates, axioms, premises, etc., are statements of the same logical standing; they differ one from another in the ways in which they have come to be formulated and the degree of confidence they enjoy. We assert a postulate and take an axiom for granted, but hypotheses we merely venture to propose. (We *believe* in hypotheses, of course, but only for the sake of argument and as an incentive to critical enquiry: "belief," as Kant defined it, "is a kind of consciously imperfect assent.") The term "premise" is coming to acquire an

[38] See *The Technique of Theory Construction* by J. H. Woodger (Univ. of Chicago Press, 1939).

antiquarian flavor. The Rev. Sydney Smith, a famous wit, was walking with a friend through the extremely narrow streets of old Edinburgh when they heard a furious altercation between two housewives from high-up windows across the street. "They can never agree," said Smith to his companion, "for they are arguing from different premises." I explain the point of this story later on. Hypotheses and axioms may be shared between cognate theories, and the logical consequences of one theory may represent the starting point —the hypotheses or assumptions—of a theory of lower level. It is in this sense that complex theories may be described as logically bonded or logically articulated structures.

Scientific reasoning, in William Whewell's view, is a constant interplay or interaction between hypotheses and the logical expectations they give rise to: there is a restless to-and-fro motion of thought, the formulation and rectification of hypotheses, until we arrive at a hypothesis which, to the best of our prevailing knowledge, will satisfactorily meet the case.

2

Before discussing the "hypothetico-deductive" scheme in greater detail, let me call attention to some of its philosophical implications.

If we accept the idea that scientific reasoning is a kind of dialogue between the possible and the actual, between what might be and what is in fact the case, then we are narrowing down the domain of science to one subclass of all possible contentions or beliefs, namely to those which are in principle capable of being modified by critical scrutiny. For, as Kant is reported to have said, it must *certainly* be true of every hypothesis that it could *possibly* be true: [39] hypotheses must be statements of such a kind that they *could* be true. During the nineteen-thirties, logical positivists used "verifiability in principle" as an operational criterion of meaning, one that served to distinguish significant statements from the various allotropic forms of nonsense.

It is not now generally believed that "verifiability" is a property that will serve this purpose. In the first place "falsifiability" should be substituted for "verifiability," as Popper has recommended.[37] The arrows in my diagram are intended to signify the polarity or one-wayness of deductive reasoning. Deduction, properly carried out, guarantees that if our hypotheses (axioms, assumptions, etc.) are true, then so also, necessarily, will be the inferences drawn from them. If therefore a hypothesis leads to expectations which are not borne out, there *must* be something wrong. But if our expectations are fulfilled, it does not by any means follow that the hypotheses which gave rise to them are true, for false hypotheses can lead to true conclusions. We may accept the hypothesis "as if it were perfectly certain," but "hypotheses always remain hypotheses, i.e. suppositions to the full certainty of which we can never attain" (Kant [39]).

In the second place (again following Popper), "falsi-

---

[39] See Kant's *An Introduction to Logic*, pp. 75-76, trans. C. K. Abbott (London, 1885). With Kant's approval, the text was compiled by a student from Kant's lecture notes. Kant's view of the nature and use of hypotheses is very much in the modern style.

fiability" marks the distinction between, on the one hand, statements that belong in science and to the world of common sense, and on the other hand statements which, though they belong to some other world of discourse, are not to be dismissed contemptuously as nonsense. Metaphysics is a compost that can nourish the growth of scientific ideas. But if we accept falsifiability as a line of demarcation, we obviously cannot accept into science any system of thought (for example, psychoanalysis) which contains a built-in antidote to disbelief: to discredit psychoanalysis is an aberration of thought which calls for psychoanalytical treatment. (The critic cannot win against such a contention—but he does not have to compete.)

As Whewell pointed out (though Mill, rather perversely I think, disagreed with him),[40] the extramural implications of a hypothesis are often specially valuable —the new expectations it gives rise to, lying outside the phenomena it was originally intended to explain—because they make it possible to expose the hypothesis to independent or at least to unpremeditated tests. "Predictions" of this kind, when they come off, are a source of special satisfaction. Yet it should not be thought that a good hypothesis is one that explains everything it is applied to. The smooth facility of Freudian and the older evolutionary formulae [41] exasperated those who

[40] See *The Art of the Soluble,* pp. 141, 149. For the argument between Mill and Whewell, see Whewell, *The Philosophy of the Inductive Sciences* (2nd ed.), pp. 62-64; Mill, *System* (later editions), Book III, chap. XIV, § 6; Whewell, *Philosophy of Discovery,* pp. 272-274.

[41] For the shortcomings of older Darwinian formulae, see *The Art of the Soluble,* p. 41.

were dissatisfied with them because, though explaining everything in general, they explained nothing in particular. Popper takes a scientifically realistic view: it is the daring, risky hypothesis, the hypothesis that might so easily not be true, that gives us special confidence if it stands up to critical examination.

3

How well does the "hypothetico-deductive" scheme of thought measure up to the specifications of a good methodology, as I tried to lay them down at the end of my second lecture?

1. A clear distinction is made between discovery and justification or proof,[42] now resolved into two separate and dissociable episodes of thought.

2. The initiative for the kind of action that is distinctively scientific is held to come, not from the apprehension of "facts," but from an imaginative preconception of what might be true.

3. The hypothetico-deductive scheme provides a theory of special incentive. Our observations no longer range over the universe of observables: they are confined to those that have a bearing on the hypothesis under investigation.

4. It allows also for the continual rectification or running adjustment of hypotheses by the process of negative feedback I shall mention again below.

[42] "Proof" in the sense of probation, the act or action of testing or trial. "Proof" in a logical context usually refers to a test of which the outcome is verification: proving a theorem is not just testing it, but testing it and finding it to be true.

5. Error is simply explained—the fact that scientific research so very often goes wrong. Scientific error is now an ordinary part of human fallibility: we simply guess wrong, take a wrong view, form a mistaken opinion.

6. Luck, unintelligible in inductive reasoning, now makes sense. The lucky accident fulfills a prior expectation, however vaguely formulated it may have been.[43]

7. The hypothetico-deductive scheme gives due weight to the critical purposes of experimentation: we carry out experiments more often to discriminate between possibilities than to enlarge the stockpile of factual information.

So far all is well. Now let us turn to certain shortcomings, real or fancied, of the hypothetico-deductive scheme.

If it is a formal objection to classical inductivism that it sets no upper limit to the amount of factual information we should assemble, so also is it a defect of the

[43] We scientists often miss things that are "staring us in the face" because they do not enter into our conception of what might be true, or, alternatively, because of a mistaken belief that they could not be true (see below, note 44). In our earlier work on immunological tolerance (*Philos. Trans. Roy. Soc.,* B, **239** [1956]: pp. 357-414), my colleagues R. E. Billingham and L. Brent and I completely missed the significance of observations which, rightly construed, would have led us to recognize an altogether new variant of the immunological response (the "graft against host" reaction) which now plays a very important part in the theory of tissue transplantation. The "facts" were before us, and if induction really worked we should not have been obliged to wait several years for their elucidation, which was hit upon independently by M. Simonsen and by Billingham and Brent themselves (see Billingham's Harvey Lecture on *The biology of graft-versus-host reactions.* Harvey Lectures, Series 62, New York, 1968).

hypothetico-deductive scheme that it sets no upper
limit to the number of hypotheses we might propound
to account for our observations. To exchange Whewell's
system for Mill's is, on the face of it, to trade in an infinitude of irrelevant facts for an infinitude of inane
hypotheses. Mill meant it as a criticism, not as a comment, when he said (*System,* Book III, Chap. XIV
§ 4):

> An hypothesis being a mere supposition, there are no other
> limits to hypotheses than those of the human imagination; we may,
> if we please, imagine, by way of accounting for an effect, some
> cause of a kind utterly unknown, and acting according to a law
> altogether fictitious.

In real life, of course, just as the crudest inductive observations will always be limited by some unspoken
criterion of relevance, so also the hypotheses that enter
our minds will as a rule be plausible and not, as in theory
they could be, idiotic. But this implies the existence
of some internal censorship which restricts hypotheses
to those that are not absurd, and the internal circuitry
of this process is quite unknown. The critical process
in scientific reasoning is not therefore wholly logical in
character, though it can be made to appear so when we
look back upon a completed episode of thought.

A second objection is this: that although falsifiability
is a logically conclusive process—if our inferences are
false, the axioms from which we deduced them must be
false also—we may yet be fallible in our imputation of
fallibility. We could be mistaken in thinking that our
observations falsified a hypothesis: the observations
may themselves have been faulty, or may have been
made against a background of misconceptions; or our

experiments may have been ill-designed. The act of falsification is not immune to human error.[44]

My third point is comment, not criticism. There is nothing distinctively scientific about the hypothetico-deductive process. It is not even distinctively intellectual. It is merely a scientific context for a much more general stratagem that underlies almost all regulative processes or processes of continuous control, namely *feedback,* the control of performance by the consequences of the act performed. In the hypothetico-deductive scheme the inferences we draw from a hypothesis are, in a sense, its logical output. If they are true, the hypothesis need not be altered, but correction is obligatory if they are false. The continuous feedback from inference to hypothesis is implicit in Whewell's account of scientific method; he would not have dissented from the view that scientific behavior can be classified

[44] The force of this objection was impressed upon me by Dr. Ernest Nagel. I can remember forming the opinion, as a young research worker, that cells long maintained in culture outside the body undergo a transformation into the cancerous state, but dropped it because of the mistaken belief that transplantation experiments had already proved the idea untenable. My reasoning was that, if long-cultivated cells were in fact malignant (as many are now known to be), they should grow progressively, as malignant tumors do, when reimplanted into the body. In fact they did not do so, so the hypothesis was disproved. Unfortunately the act of disproof was itself erroneous: we now know that such a test could only have been valid if the cultivated cells contained no transplantation antigens not also present in the organism into which they were implanted; and even this test would have been unreliable if the cultivated cells had acquired new tumor-specific antigens during their growth outside the body. It is indeed true that ". . . disproof of a hypothesis is contingent on the stability of the theories employed in interpreting matters of fact, so that the refutation of a supposed explanation may be no more definitive than is its verification" (E. Nagel, in a review of *The Art of the Soluble, Encounter,* September, 1967: pp. 68-70).

as appropriately under cybernetics as under logic.[45]

The major defect of the hypothetico-deductive scheme, considered as a formulary of scientific behavior, is its disavowal of any competence to speak about the generative act in scientific enquiry, "having an idea," for this represents the imaginative or logically unscripted episode in scientific thinking, the part that lies outside logic. The objection is all the more grave because an imaginative or inspirational process enters into *all* scientific reasoning at every level: it is not confined to "great" discoveries, as the more simple-minded inductivists have supposed.

Scientists are usually too proud or too shy to speak about creativity and "creative imagination"; they feel it to be incompatible with their conception of themselves as "men of facts" and rigorous inductive judgments. The role of creativity has always been acknowledged by inventors, because inventors are often simple unpretentious people who do not give themselves airs, whose education has not been dignified by courses on scientific method. Inventors speak unaffectedly about brain waves and inspirations: and what, after all, is a mechanical invention if not a solid hypothesis, the literal embodiment of a belief or opinion of which mechanical working is the test?

Intuition takes many different forms in science and mathematics, though all forms of it have certain prop-

[45] "Trial and error" will not do as a description of the process by which we devise and test hypotheses, for it carries the sense of random exploration, or of exploration according to a scheme ("let's try the following possibilities in turn") which is not influenced by the testimony of prior mistakes.

erties in common: the suddenness of their origin, the wholeness of the conception they embody, and the absence of conscious premeditation. The four examples I shall give are not meant to be exhaustive or mutually exclusive.

(*a*) *Deductive intuition:* perceiving logical implications instantly; seeing *at once* what follows from holding certain views. Inasmuch as deductive reasoning merely uncovers or brings to light what is implicit in our premises, a pedant might insist that deductive intuition is not a "creative" process. To a perfect mind Pythagoras' Theorem (the one that startled Thomas Hobbes) would simply be a boring and repetitious way of underlining a point which had already been made, much more compendiously, in Euclid's axioms.[46] It is indeed true that deduction owes its existence to the infirmity of our powers of reasoning: it cannot bring us news of the world, but (because our minds are indeed imperfect) it can bring us awareness.

(*b*) The form of intuition which, unless we are to abandon the word altogether, might as well be called *inductive:* thinking up or hitting on a hypothesis from which whatever we may wish to explain will follow logically.[47] This is the generative act in scientific discovery, the invention of a fragment of a possible world. "Creativity" is a vague word, but it is in just such a context as this that we should choose to use it.

[46] See for example Hans Hahn's influential paper on "Logic, Mathematics and Knowledge of Nature" (1933), reprinted in *Logical Positivism,* ed. A. J. Ayer (New York, 1959).

[47] It is a grievous blunder to speak, as so many do, of *deducing* hypotheses. Hypotheses are what we deduce things from.

(*c*) The instant apprehension of analogy, i.e. a real or apparent structural similarity between two or more schemes of ideas, regardless of what the ideas are about. No one word in common use describes this faculty in all its manifestations, but if I had to choose one word I should choose *wit* (*cp.* the anecdote about Sydney Smith, p. 48 above).

(*d*) Most scientists cannot be classified as either experimentalists or theorists, because most of us are both, but we all recognize a distinction between the faculties that found superlative expression in, say, Michael Faraday on the one hand or James Clerk Maxwell on the other. For an experimentalist the most exciting and pleasing act in science is thinking up or thinking out an experiment which provides a really searching test of a hypothesis. We recognize the intuitive element in such a process when we speak of experimental flair or insight, but here too no one word in common speech stands for everything it should convey.

The analysis of creativity in all its forms is beyond the competence of any one accepted discipline. It requires a consortium of the talents: psychologists, biologists, philosophers, computer scientists, artists and poets will all expect to have their say. That "creativity" is beyond analysis is a romantic illusion we must now outgrow. It cannot be learned perhaps, but it can certainly be encouraged and abetted. We can put ourselves in the way of having ideas, by reading and discussion and by acquiring the habit of reflection, guided by the familiar principle that we are not likely to find answers to questions not yet formulated in the mind. I am not offended by the idea that drugs may help us to formulate hypotheses, but I know of none which improves their quality, and I should hesitate to use a drug which did not enhance the critical faculty in proportion to the rate of accession of ideas.

4

The scheme of thought I have outlined in this third lecture explains the balance of faculties that should be cultivated in scientific research. Imaginativeness and a critical temper are both necessary at all times, but neither is sufficient. The most imaginative scientists are by no means the most effective; at their worst, uncensored, they are cranks. Nor are the most critically minded. The man notorious for his dismissive criticisms, strenuous in the pursuit of error, is often unproductive, as if he had scared himself out of his own wits—unless indeed his critical cast of mind was the consequence rather than the cause of his infertility.

The hypothetico-deductive system seems to me to give a reasonably lifelike picture of scientific enquiry, considered as a form of human behavior. It makes science very human in its successes as well as in its fallibility. "Let us look each other in the face," said Nietzsche:[48] "we are Hyperboreans: we know well enough how far outside the crowd we stand." Nietzsche was speaking about philosophers, to be sure, and more par-

[48] *The Antichrist,* trans. A. M. Ludovici (London, 1911). The Hyperboreans, inhabitants of a serene and timeless world lying beyond the roots of Boreas, the North Wind, seem to owe their existence to a clerical error: "Hyperphoreans," I have been told, may be a sounder rendering of their original name, so they are doubly mythical creatures.

ticularly about himself; but philosophers have long since outgrown the hyperborean image, and (to whatever degree it may have been wished upon him) the scientist must outgrow it too. The scientific method is a potentiation of common sense, exercised with a specially firm determination not to persist in error if any exertion of hand or mind can deliver us from it. Like other exploratory processes, it can be resolved into a dialogue between fact and fancy, the actual and the possible; between what could be true and what is in fact the case. The purpose of scientific enquiry is not to compile an inventory of factual information, nor to build up a totalitarian world picture of natural Laws in which every event that is not compulsory is forbidden. We should think of it rather as a logically articulated structure of justifiable beliefs about nature. It begins as a story about a Possible World—a story which we invent and criticize and modify as we go along, so that it ends by being, as nearly as we can make it, a story about real life.

INDEX